The toys wanted to plan a picnic.

"I'll bring food to eat," said the truck.

"I'll heat beans," said the toy stove.

"I'll boil the tea," said the toy kettle.

"Ring, ring, please start," said the bike.

Go ahead! Enjoy each bite!

Jack-in-the-box will read a book.

What a great picnic!